# Ham Radio Reference:

Easy Ham Radio Guide For Beginners And Advanced

**Disclamer:** All photos used in this book, including the cover photo were made available under a Attribution-NonCommercial-ShareAlike 2.0 Generic and sourced from Flickr

Table of Contents

Ham Radio Reference: ...................................................................................1
    Easy Ham Radio Guide For Beginners And Advanced..............................1
Introduction ................................................................................................3
Chapter 1 – Basic and Advance Concepts of Ham Radio ..........................5
Chapter 2 – Tips and Tricks to Learn Morse Code ....................................9
Chapter 3 – Assemble Your Own Station ................................................14
Chapter 4 – What is DXing?......................................................................20
Chapter 5 – Common Mistakes in Operations with Ham Radio ...............23
Conclusion ................................................................................................27

## Introduction

With the advent of modern communication tools, the use of Ham Radios has seen a decline. Moreover, questions regarding the credibility of these radios have been raised with the passage of time. In order to prove the efficiency of these radios, one must consider their ability to work at the times of crises. In the numerous events of the history, where all the other advanced tools had stopped working, these Ham Radios were proved beneficial.

The Ham Radios possess multiple advantages. Other than their ability of efficient communication during emergencies, it is also used for shortwave communication. In addition, where other radios might demand batteries or grids for recharging, these radios only demand a few power supplies. Their power requirements can be fulfilled using solar or even hydro resources.

Henceforth, the chapters will incorporate the basic concepts to familiarize the user with a Ham Radio. The concepts inculcate all of those steps which must be followed in order to learn the major techniques required to run a Ham Radio. In addition, the tips and techniques, which are used to learn the

Morse code, are also encompassed here. Morse code is actually used to represent the alphabets in the form of signals. Thus, it is important to learn the advance encoding techniques used for the communication purposes.

This book includes a detailed analysis regarding the mode of communication in the ham radios i.e., Morse code. These codes are composed of ditz and dahs.

These ditz and dahs are the codes to encrypt the signals. After complete encryption of a word, space is added using the same coding procedure. Therefore, these codes are extremely useful.

In addition, the types of the ham radios along with the frequencies used are also included to facilitate the user. Hams are classified on the basis on the basis of their size, shape, and their respective frequencies. The procedure to assemble your own radio station is included as well.

Another interesting piece of information included in the book is DXing. DXing means listening to the radio channels of foreign areas. These stations are out of the usual range. However, many radio operators, who are interested in unique channels, use this technique for enjoyment. All of those persons, which have the hobby of DXing are known as DXers.

Therefore, a Ham Radio is a communication tool, which connects people, communication and electronic altogether. These radios are used to connect people around the globe, within a few minutes. Although, a large group considers these radios as old-fashioned, yet there is no denial to the fact that these radios are one of the best communication tools even in the modern day world.

## Chapter 1 – Basic and Advance Concepts of Ham Radio

Ham radio, also known as Amateur radio, is a useful tool designed for the purpose of non-commercial communication along with the wireless experimentation. The name, amateur, is used in order to distinguish it from the commercial communication.

Moreover, the operator of this radio is known as a ham, which is licensed by the Federal Communications Commissions. In addition, all of the hams must have the basic knowledge about the operating principles, so that they can pass the examination.

Thus, in order to provide detailed information regarding the basic ideas of a ham radio, consider the following guidelines:

1. **Public security of a ham radio:**

    Firstly, the ham radios are one of the most important means of communications that works in adverse situations as well. It is because of their incredible robust mode, which increases their importance. Ham radios have proved themselves in hurricanes as well as in other crisis. Even the bad weather cannot stop these radios from working and ensuring communication. These radios are used to provide weather updates and traffic situations in severe weather conditions.

2. **Range of a ham radio:**

A ham radio possesses a wide range of frequencies. You can tune to the frequency of your own choice and tune into the radio spectrum, which enables the user to enjoy. In order to explain this concept, consider that a ham radio is a long measuring tape. Starting from 1/2 inches and extending towards 1.7 inches, is the AM radio band. After this, the FM radio band begins. Here the measuring tape analogy is used to explain the working condition of a radio. There is a keen division in this frequency range. You can choose the frequency depending upon your requirement.

### 3. How does a ham radio work in mountainous regions?

Sometimes, when the radio waves gain the line of sight path above 30 MHz, then difficulty might arise in communication. In order to overcome this barrier, several repeaters are attached at the top of mountains and tall buildings. The network of these repeaters is very sophisticated and efficient. Sometimes, the network is designed in such a manner that the repeater goes to the internet, which is then used to pop out some other repeater of the network. Therefore, the problems in the communication in mountainous areas are resolved.

### 4. The procedure to get Ham license:

Ham license is given after the test is passed. The test is not difficult at all. You can learn the new skill, within a day or two. After you have mastered the skills, then you can apply for the license. In order to learn about the ham radio, start browsing. To begin, download the manuals regarding the working of a ham radio.

## Types of ham radios:

Ham radios possess various types. They are differentiated on the basis of their size, mode of communication and numerous other parameters. However, following are three common types of ham radios:

### 1. Handheld ham radios:

Handheld radios are small and light in weight. These radios have a typical range of 5 to 7 miles. Furthermore, they produce low power output. Moreover, their range can be increased by using a repeater.

### 2. Mobile ham radios:

These comparatively bigger radios are used for operational purposes. In addition, since they possess higher power output, their range is also higher i.e., 200 miles.

### 3. Base- Station ham radios:

These kinds of radios are the largest of all kinds. They of radios possess a large frequency spectrum. In addition, because of their higher battery output they possess a higher range.

## Frequencies of a ham radio:

Interestingly, ham radios possess a wide range of frequency spectrum. Therefore, it is quintessential to learn about the frequencies of ham radio. It possesses the following kind of frequencies:

1. **High Frequency (HF) - 3 MHz to 300 MHz:**

   This higher frequency range is usually used by a large number of hams to provide powerful broadcast of the channels. Moreover, it is also used because it ensures better communication.

2. **Very High Frequency (VHF)- 30 MHz to 300 MHz:**

   Some of the transceivers work only on higher frequencies. Therefore, these frequencies are also used in many ham radios.

## Chapter 2 – Tips and Tricks to Learn Morse Code

In 1884, F.B. Morse, developed the Morse code for efficient communication. It is because of the simplicity yet efficaciousness that these Morse codes are used even today. Morse codes are composed of ditz and dahs, which are used by a number of people for communication purposes.

Moreover, the use of these codes is not limited to simple communication. It is also used for encrypted communication for security purposes. This code is used for Emergency Signaling (SOS) for all kinds of receivers. This code can also be used by the people with several disabilities.

The Morse code is mostly used by the hams because they ensure systematic transmission of the signals. However, it is not very easy to learn these codes. In order to get better results, one has to learn them as a new language. Following are the steps, which will help in learning the tips and tricks of the Morse code:

1. **Listen to the Morse code recordings:**

   When you began listening to the Morse codes, you start sensing that they have a certain combination of dashes and dots. These dashes and dots are also known as the dahs and ditz. Interestingly, the dahs are the comparatively longer beeps, whereas, the ditz are the shorter beeps. Each alphabet of the word is separated by a shorter pause. Every word, on the other hand, is spaced by a longer pause. In order to measure the speed of Morse code, words per minutes are used.

   One can buy these recordings online. Moreover, shortwave receivers are also used to listen to these recordings. In addition, there are several training applications available on the internet, which can prove to be more beneficial as compared to these recordings. The software used, provides numerous ways to teach Morse codes. Thus, enabling you to choose the method with which you are comfortable. For better learning, try to learn the sound of an alphabet rather than counting the ditz and dahs.

2. **Use a Morse code chart:**

   Morse codes made up of ditz and dahs are used as a mode of communication. These codes include the encryption of alphabets to ensure safe transmission of the signals. Moreover, these codes can be learned using a chart and then they create the sounds. Although these codes are outdated, yet they are still used. It is because of their efficiency and strength that they are used for secure communication. Furthermore, these codes are used in order to help in the communication of the disabled individuals.

One of the methods to learn Morse code is by carefully analyzing the chart for Morse code alphabet. These charts include punctuation, Q codes, and abbreviated phrases along with the ditz and the dahs arrangement for various words. For better results, practice them a lot. Try to learn from your mistakes.

Some people write down the alphabets with ditz and dahs and then improve their knowledge of the Morse code. Some, on the other hand, try to adopt innovative ways to skip this additional step. Furthermore, some charts that list the sound of these Morse code signals are also used by many people. The key here is to learn by that method, which you find easy.

3. **Begin with simple translations:**

In the initial stages, one must start writing these codes and then continue with creating their sounds. However, once you have practiced, you must start making sounds of the basic words. In order to learn the pronunciation of the Morse codes, bear the fact in your mind that dit is pronounced as di, where the t is silent and a short i sound. Dah, on the other hand, is pronounced as a shorter sound.

Once you have learned these basic techniques, then start practicing on simple and easy words. However, while practicing, you must not write the codes rather make a sound of them. In order to check your results, record the sound you have created.

However, care must be taken while adding spacing. If your spacing is appropriate, then your code will be easier to understand.

### 4. Start memorizing the simple alphabets:

Another useful tip while learning Morse code is to begin memorizing with the alphabets that have easier Morse codes. For instance, T is represented by a single dah. E, on the other hand, is depicted using a single dit. Moreover, dah-dah is used to represent an M, whereas, the dit-dit is used for I.

After you have learnt these codes, then jump on to learn the combinations. Fortunately, the alphabets with long Morse codes are not used commonly. Some of the alphabets with long Morse codes are Q, K, Z, V and X.

### 5. Make associations:

Mark some similarities to learn the codes. For instance, C has a Morse code of dah-dit-dah-dit. You can remember this code by considering that the code follows the sequence of long-short-long-short.

There are other alphabets as well, which can be memorized using similar mnemonics. Moreover, you can also associate the sound of these alphabets with any of your favorite melodies.

Therefore, it is recommended that if you want to learn the Morse codes, and then try to recognize some of the associations between the code with any melody or pattern.

Therefore, Morse code is used universally for communication purpose. As it is an old yet very useful mode, therefore, it is used even today for transmitting the signals to other locations.

## Chapter 3 – Assemble Your Own Station

Usually some high frequency radios stations come in the mind whenever, one thinks of a ham radio. However, these radios are used for communication purpose throughout the world. They possess several kinds depending upon the receivers, the location and the requirement.

Henceforth, we will focus on the major techniques that are used to set up a common Ham radio. However, one must bear in mind that you can only use these radios after you have got the license. Given below is the procedure to assemble your own station.

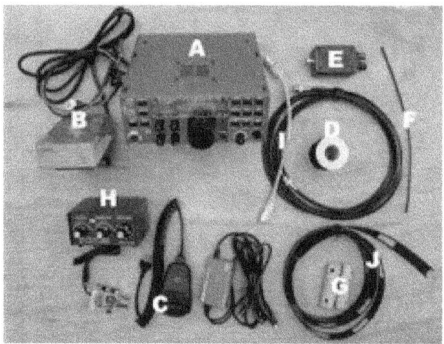

Selection of the appropriate equipment is one of the most critical parts in the installation of a new radio station. Following is the list of recommended equipment that must be used in the installation.

1. High Frequency radio(a):

Radio forms the basis of any station; therefore, they must be chosen with appropriate care. You must choose the radio which can transmit as well as receive the signals on the ham radio. A HF radio must be multi-mode with a hundred Watt power source.

The multi-mode radio is the one which supports all kinds of communication i.e., digital, Morse code and voice communication. Moreover, a multi band radio is yet another suitable characteristic. A multi-band radio works with a wide spectrum of frequencies.

## 2. Power source(b):

Usually the operating power of almost all kinds of ham radios starts from 13.8 V DC. The switches at houses provide 120 V AC; this implies that the radio might need another source of power supply.

A switch mode supply is recommended, as it is a better power source than a linear one. Astron SS-25 can be used for this purpose. Moreover, other products can also be used, which provide you the current and the voltage reading on the front.

## 3. Accessories(c):

In most of the radios, a microphone is used for the purpose of voice communication. However, for Morse code, you might need a CW key. For this purpose, you can invest in an iambic key.

This will cost you around 130 dollars. In addition, TNC is used to interface your radio with the computer for other digital modes of communication. RigBlaster Plug and Play is usually recommended.

### 4. Antenna(d):

Interestingly, the most critical part of the ham radio installation is the placement of the antenna. There are numerous designs available that cause difficulty in choosing. However, while choosing an antenna, you might have to face certain barriers.

The choice of antenna depends upon your current location. Try to choose that arrangement of antenna which suits your surroundings. It must be capable of catching the signals from the repeaters and other transmitters. If you are unable to find a suitable design for your location, then you can always design one on your own.

### 5. Balun(e):

The word Balun is a combination of balanced and unbalanced. Balun is used to create balance in the communication procedure. It is used to transform an unbalanced signal into a balanced one. One of the most efficient Balun used is the LDG RBA-4:1.

### 6. Heat shrink tubing(f):

Heat shrink tubing consists of a plastic wire, which is used to isolate the wires. Moreover, they are also used for environmental safety. They are used to cover the joints and terminals. They are usually made up of poly-olefin and nylon.

### 7. Dog bone Insulators(g):

These insulators are used to connect two conducting elements. They combine them in such a way that there is no conduction between them.

### 8. Tuner(h):

LDG Z-11 Pro 2 is a completely automatic design, which is used for tuning purposes. It costs around 179 dollars. However, the manual tuners are less expensive.

### 9. Coaxial cable(i):

RG-58 can be used for wiring purposes. Moreover, RG-8X can also be used.

### 10. Ladder line(j):

Ladder line is also known as a window line. It is used to provide power to a multi-band antenna. 450 ohm window line can be used in order to install a common ham radio station.

**Tools:**

The required tools and materials can vary depending upon the arrangement you have to make. However, following are the tools, which are required for assembling a simple ham radio:

i. Solder.

ii. Embossing heat tool.

iii. Soldering iron.

iv. Wire strippers.

v. Wire cutters.

These tools and equipment are the initial steps in the installation of a ham radio. After these comes the actual implementation. In order to assemble the radio, one must follow the steps given below:

1. The foremost step is to understand the working of the dipole antenna. The dipole antenna is a thin and long wire, which is used to catch the required frequency.

2. Afterwards, you have to look out for the arrangements to fit the dipole antenna. The antenna must be arranged in such a manner that it is not affected by a storm or any other similar thing. Moreover, it must be placed in such a position that it catches appropriate signals.

3. Now, connect the antenna with the window line. After you have connected them, secure them with soldering the terminating points.

4. Once you have connected the antenna with a window line, connect the antenna with a dog bone insulators.

5. Now connect the balun, tuner and the radio in the given order.

Once you have settled the antenna, you might face some problems in the signal reception. There is no need to worry about such problems; they can be solved by fixing the antenna and the ladder line.

## Chapter 4 – What is DXing?

The X used in DXing implies the unknown distance, whereas, the D used is known for distance. This DX refers to the unknown distance. DXing is, therefore, listening to the foreign radio channels.

It refers to the act of listening foreign radio stations, which does not lie within the normal range. Thus, listening to the casual radio stations does not mean DXing. Similarly, DXers are those hams, which enjoy themselves by listening to the foreign radios.

In addition, DXers expand their activity by broadcasting several stations themselves. Some of the DXers broadcast those pieces of information, which are supposed to be a secret. Thus, they interrupt the signals of many mobile phone companies as well as the military signals and broadcast them. If such illegal activities are pursued, then DXing can be dangerous as well.

However, the major test lies in the catching of the radio signals which are present at long distances. In different parts of the world, there are still many radio stations, which are alien to a number of people.

However, these stations can be explored by an experienced DXer, who is familiar with all of the shortcuts and tips. A Dxer must possess enough skills to make interesting yet legal broadcasts.

In the previous times, DXing was considered as a unique yet interesting way to explore the people of different areas. Before the world turned into a global village, DXing was the connecting bridge.

Even in today's world, DXing is an interesting way to connect with the people of other countries. There are numerous receivers available, which are specially designed to help the hams in DXing. They can still follow this interesting hobby using these specially designed receivers. These receivers possess shortwave bands.

**Shortwave:**

Despite the fact that DXing is practiced on numerous bands, there are still a lot of DXers, which use short or medium wave for DXing. However, not all of the shortwave users desire to listen to the foreign radios. Some of these DXers might tune in to listen to their favorite show. While some, on the other hand, use these waves for DXing.

Ionosphere, a layer of Earth's atmosphere, is charged by the sun. Therefore, it possesses the property of reflecting some of the radio waves. This layer acts like a huge mirror and reflect the radio waves with certain frequencies. It must be noted that not all of the radio waves are reflected, but some with low frequencies are absorbed, while others are reflected.

The upper and lower limit of reflection and absorption keeps on varying. These limits keep on changing due to certain environmental factor. Normally the reflected waves are known as the shortwaves. It is because of this reason that the shortwaves are more commonly utilized in the radio communication.

There are numerous advantages of listening to the different stations on shortwave. It is one of the best methods for transmitting the news of the world. Additionally, shortwave is also used to learn about the cross cultural information. They are further used to share information regarding politics and events.

DXers use the unique hardware which is then used to analyze the signals. The receivers of this mode of communications are particularly intended for common audience. However, an expensive receiver is often considered as a pointless apparatus used for reception. Most DXers use reception apparatuses made of copper wire, which is held tight on trees or other towers.

Therefore, Dxing is a common method, which is used by a large number of people to listen to the radio channels, which are foreign. These channels are not used for broadcast, yet they possess extremely high importance. Moreover, more than a profession, it is a hobby that requires no license.

## Chapter 5 – Common Mistakes in Operations with Ham Radio

Ham radios are used universally for communication purposes; however, there are certain problems which have been noted in their operations. These problems are mainly caused because of some basic errors in the installation. After a series of experimentation and research, many problems have been observed. Therefore, in order to eradicate these problems, following errors must not be made:

1. **Choose the right hardware:**

    As the antennas and other hardware is generally installed in the open air; therefore, it is recommended that such hardware must be used, which is prone to corrosion. Usually the antenna made up of stainless steel or galvanized materials is used to avoid corrosion.

    Moreover, the use of wrong hardware is another common problem. A lot of people try to replace expensive, yet reasonable hardware with the inexpensive one.

    This must not be done at any cost. Hence, it is better to buy expensive hardware once rather than buying the inexpensive one multiple times. Also, try to use the hardware that is in correspondence with your surroundings.

2. **Try to join the tower sections on the ground:**

   Another issue that is usually faced with the installation of a ham radio is fitting multiple sections on the tower, instead of the ground. Hence, it is recommended that the tower sections, no matter new or used, must be connected on the ground before their installation on the tower. In addition, tower*jack combination is used to align these sections together.

3. **Lack of care:**

   The entire ham radio apparatus is constantly under the state of deterioration. Thus, it is recommended to check the fittings annually. Moreover, one of the best ways for this is to solve the small problems first before they become gigantic issues. Moreover, the nuts and bolts of the tower must be checked constantly after a short period of time.

4. **Absence of a proper ground system:**

   Another issue, which is faced after the installation of a ham radio, is the absence of a proper grounding system. Ground system is essential in order to provide a zero level to the voltage flowing from the higher level.

   The ground system is quintessential to protect the entire radio station from lightening and minimizing the RFI to the nearby electronic wires or systems. This is very common and huge mistake; therefore, one must try to avoid it at all costs.

### 5. Inappropriate wiring:

Inefficient wiring is another problem that causes severe problems. If the coaxial cables are not insulated properly, then transmission losses can occur. Moreover, the absence of insulation results in naked wires, which are extremely dangerous if touched by any human or even an animal. These wires can even cause deaths as well as fatalities.

### 6. Underestimating the wind forces:

Antenna and towers can be affected by the wind pressure to a great extent. Unless you have experienced the pressure yourself, you cannot measure the impacts the storm or pressure can cause.

Even a small increment in the wind speed can increase the pressure to a tremendous amount. Moreover, the wind pressure keeps on changing in an abrupt manner; thus, it is recommended that one must install the hardware to mitigate any crisis caused by the wind pressure.

### 7. Follow the manufacturer's guide:

Whenever you buy a ham radio station, you are provided with a manual, which is written to ensure an appropriate installation. Moreover, this manufacturer's guide also includes the precautionary measure to mitigate any unusual behavior.

After a series of research, it is observed that many towers and radio stations fail to work because they lack the basic details. If these details are fulfilled, then these issues can be eradicated.

## 8. Overloading:

Overloading is one of the most common reasons that cause the failure of ham radio towers. The foremost thing you must consider before installing any tower is the wind speed rating. Afterwards, you must also consider the specifications of the hardware as per the guide provided by the manufacturer.

It is an undeniable fact that a lot hams overload the receiver and antennas, which results in severe problems. Additionally, sometimes the engineering specifications of the tower are not efficient enough to carry the burden of the signals. Therefore, it is not a good idea to overload your antenna.

## 9. Inefficient power supply:

The ham radio stations need a power supply in DC, whereas the switch power is in the AC form. A large number of people try to provide a power source to the radio via switch. Because of the compatibility issues, the radio fails to work. At that point the need of an appropriate power supply arises. Hence it is recommended that suitable power supply must be guaranteed.

## Conclusion

Ham radios are considered as one of the oldest yet efficient mode of communication. However with the passage of time, various questions have been raised regarding their efficiency.

Therefore, in order to prove their practicality, this book provides the details, which prove the proficiency of the ham radios. Their ability to work under the crisis has enhanced their value to many fold. History depicts that during the times of need, these radios have rescued us, whereas, other modern day technologies have failed to provide any help.

The Ham Radios have numerous benefits. To begin, ham radios are not only used in a crisis situation, but also, they are used for shortwave communication. In addition, where different radios demand batteries or huge power supplies, these radios just request a small amount of power. Interestingly, they can work on solar cells, hydro-electric cells, and windmills.

In the various chapters of this book, the fundamental ideas to assemble a ham radio are included. The ideas instill those strides which must be followed to ensure the efficacious working of the radio station. Moreover, the tips and procedures, which are utilized to use the Morse code, are likewise included. Morse code is used to transmit signals using various dashes and dots.

This book incorporates a detailed analysis regarding the method of communication of the ham radios i.e., Morse code. These codes are made out of ditz and dahs. These ditz and dahs are the codes to encrypt the signs. After complete encryption of a word, space is included utilizing the same coding system. In this manner, these codes are to a great degree valuable.

In addition, the different kinds of the ham radios alongside the frequencies utilized are also included to provide benefit to the reader. Hams are arranged on the premise of their size, shape, and their particular frequencies. The strategy to amass your own particular radio station is incorporated also.

Another fascinating piece of information incorporated into the book is DXing. DXing implies listening to the radio channels of remote regions. These stations are out of the standard extent. Notwithstanding, numerous radio administrators, who are keen on exploring new channels, utilize this system for delight. Those persons, which love to explore new channels using DXing are known as DXers.

Thus, a Ham Radio is a specialized apparatus, which is used to connect individuals. These radios are utilized to associate individuals around the world, within a few minutes. In spite of the fact that, a vast gathering considers these radios as antiquated, yet they are one of the most reliable modes of communication.

### FREE Bonus Reminder

If you have not grabbed it yet, please go ahead and download your special bonus report *"Preppers Survival Guide. Proven Tactics For Armed Incounters!"*

Simply Click the Button Below

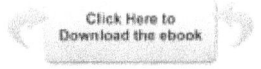

OR **Go to This Page**

http://preppersliving.com/free

## BONUS #2: More Free & Discounted Books & Products

### Do you want to receive more Free/Discounted Books or Products?

We have a mailing list where we send out our new Books or Products when they go free or with a discount on Amazon. Click on the link below to sign up for Free & Discount Book & Product Promotions.

**=> Sign Up for Free & Discount Book & Product Promotions <=**

OR Go to this URL

http://zbit.ly/1WBb1Ek

www.ingramcontent.com/pod-product-compliance
Lightning Source LLC
Chambersburg PA
CBHW030041230526
45472CB00002B/612